Printed
For
Quixote Press
by
BRENNAN PRINTING
100 Main Street
Deep River, Iowa 52222
515-595-2000

A FIELD GUIDE TO ILLINOIS' CRITTERS

by

Bruce Carlson

QUIXOTE PRESS
R.R. #4, Box 33B
Blvd. Station
Sioux City, Iowa 51109

* * * * * * * * * * *

QUIXOTE
PRESS
Bruce Carlson
R.R. #4, *Box 33B*
Blvd. Station
Sioux City, Iowa
51109

PRINTED
IN
U.S.A.

Any similarity between the names of persons referenced herein and the names of actual persons, living or dead, is purely coincidental.

DEDICATION

This book is dedicated to the Monarch Butterfly. Anyone that goes through what that critter does ought to have something going for it.

TABLE OF CONTENTS

DEDICATION v
FOREWARNED ix
PREFACE xi
CHAPTERS
I SAILER CATS 13
II MAYFLIES 17
III THE PLYWOOD CUTOUT 19
IV THE-SEVEN-YEAR-ITCH 21
V NO-SEE-UMS 23
VI THE BULLHEAD 27
VII RATCHETY COONS 29
VIII DOUBLE-BREASTED MATTRESS FLAPPER 31
IX THREE-BEAN SALAD 33
X FUZZY-CHINNED-LOOP-SCOOPER 37
XI THE ILLINOIS HIGH SCHOOL BIOLOGY CLASS DISSECTION FROG 41
XII SEVENTEEN-YEAR-LOCUST 45
XIII FAT-OF-THE-LAND 47
XIV MONARCH BUTTERFLIES 51
XV THE CHEERLEADER 55
XVI THE NUMBERED COW 59
XVII BUNNIES 63
XVIII THE OPTIC 67
XIX THE STRONG-TOED BAT 71
XX GNATS 73
XXI SIRENS 77
XXII SHELF BRACKETS 81
XXIII THE AMOEBA 83
XXIV GEORGE CANE'S SEA MONSTER 85
XXV THE RED-BEAKED BESPANGLED PUSHOVER 95
EPILOGUE 99
INDEX 107

FOREWARNED

Bruce Carlson takes a fresh look at some of the animals that we've kind of learned to take for granted here in Illinois. A couple of these animals sound suspiciously like people to me, but maybe he's scared up a couple of critters out of the Illinois bush that I haven't seen yet.

Professor Phil Hey
Briar Cliff College
Sioux City, Iowa

The reader must appreciate the fact that none of these accounts have ever been published before. Some of them could cause embarrassment to living people today. Because of that, some of the accounts use fictitious names. In those cases, it should be understood that any similarity between those names and actual people, living or dead, is purely coincidental.

PREFACE

This book kind of stumbles around in the general direction of talking about some of the critters here in Illinois. A lot of it is true and educational. Some of it is kind of tongue-in-cheek, and there's a little bit of tongue-way-out-of-town.

In a couple of places I pontificate a bit on an issue or two, but not very strongly. I'm not especially good in the pontificating department.

CHAPTER I

SAILER CATS

Sailer cats are certainly not unique to Illinois, but they are one of our commonly seen furry friends. So it is important to include the sailer cat here in this book about Illinois animals.

Sailer cats are very much like other subspecies of cats here in Illinois. But in those characteristics in which they differ from other cats, they do so rather dramatically.

Their shape is a good example. Sailer cats are usually in the order of a foot and a half long, and maybe ten to twelve inches wide. Their height, however, is characteriscially only about one-fourth to one-half inch.

All members of this subspecies of cat are dead and are all found lying on the pavement of Illinois' highways.

Their name "sailer cats" comes from the Greek word "sail", meaning to fly or to sail.

If you grasp a sailer cat firmly between your thumb and forefinger as you would the lid off a coffee can, it will sail a good piece if you give it a good toss.

A word of caution, however . . . . . . we all know how hesitant Illinois lawmakers are to intrude into the lives on their citizens, but they do make an exception with sailer cats. There are rather rigidly enforced laws governing the sailing of sailer cats on public highways. One is well advised to be aware of those laws and ordinances before sailing those cute little things over into a pasture or a cornfield. It would be unfortunate to encounter a situation such as that Gary Needam from over in Rockford did when he was caught pitching a sailer cat into the Rock River. His punishment involved having to row around out there until he found the critter to bring it back.

CHAPTER II

MAYFLIES

What has no mouth, no brain, lives only twenty-four hours, and smells badly?

No, you're wrong. Politicans live more than twenty-four hours and they do have a mouth.

It's the Mayfly. You might have also heard of this charming little critter under some other name that folks come up with that is designed to insult some particular group of people.

These creatures are about an inch and a half in overall length, about that wide, and will weigh in at about .0015 pounds. Those that inadvertently fly into your face will be about a foot long and weigh about eight pounds, however.

The primary function of these river-borne insects is to give the local Chambers of Commerce something to lie about.

A real quick and dirty survey of Chambers of Commerce along mayfly-infested rivers here in Illinois has shown that they have essentially four positions on the mayfly. These are:

(1) The Mayfly does not exist.
(2) The Mayfly never has existed.
(3) The Mayfly never will exist.
(4) They are considerably worse in the next town down the river they they are in their town.

CHAPTER III

THE PLYWOOD CUTOUT

Somebody said one time; "If I can see furthur and better than those who went before, it is because I stand on their shoulders to do so.", or something to that effect.

Whoever said it, that statement speaks so well of how we build our lives and our experiences on the efforts of those who went before.

And we in Illinois can take great pride in our accomplishments with those plywood cut-out figures of folks leaning over in the garden wherein all you can see is an ample bottom and some fat legs.

There were those who spoke of bad taste reaching its zenith with the pink flamingo in the 1950s. Yet, in our own quiet way we, the children of those with that incredably bad taste, have been able to plummet to new depths with those plywood cut-outs.

The relevance of these cut-outs to this book is, of course, that those figures must be some kind of animal. After all, if they weren't some kind of critter, how could they be so much smarter than those who display them?

CHAPTER IV

THE SEVEN-YEAR-ITCH

The Seven-Year-Itch is a critter common to several Midwestern states and is certainly no more prevalent in Illinois than in any other area.

The Seven-Year-Itch is closely related to the Seventeen-Year-Locust, except it has six fewer letters and is generally thought to be a lot more fun.

CHAPTER V

NO-SEE-UMS

The official Illinois No-See-Um is not only unseen, but it is generally pretty well misunderstood, as well. The No-See-Um is often thought to be an itsy-bitsy bug that bites you on the ankle when you stomp around in grass of an evening.

That, however, is only part of the picture. The No-See-Um, like many of Illinois' life forms, has three stages. It starts out in a larval stage. It then rather quickly ~~metamerphosis~~ ~~metaphermoisis~~ ~~metamorepho~~ . . . . changes into a pupae stage, and finally emerges again as an adult. And, like any animal that undergoes such changes, each stage is dramatically different than the others.

In the larval stage of its existence, the No-See-Um is, indeed, microscopic. And it does frequent damp and grassy areas. It's not like getting nipped by a dog or even a mouse, or anything like that. It's just a crummy little irritation you have to put up with.

Besides getting around the ankles, the No-See-Um likes to get you in places hard to reach like under casts. It is also partial to anyplace that you'd rather not itch when you're out among 'em.

What's really tough is when they bite you under a cast in a place you'd rather not itch when you're out among 'em. You know you have problems then.

In a few weeks after starting out in a larval stage, the No-See-Um changes into the pupae stage and makes its presence known than by leaving little footie prints on freshly washed cars locked in garages where there is no way a No-See-Um could get in. Yet, there they are the next morning, those little footie prints across the hood or the top. Again, however, the critter is good to his name, you never actually see him.

In the adult stage, the No-see-um makes a practice out of tipping over garbage cans at night when most law-abiding critters are asleep.

CHAPTER VI

THE BULLHEAD

The bullhead that lives in Illinois' ponds and rivers is about as good a eatin' critter as there is in the Land of Bullheads.

The primary function of the bullhead is, however, not to be eaten, but to provide the humans in Illinois with a sense of their own limitations.

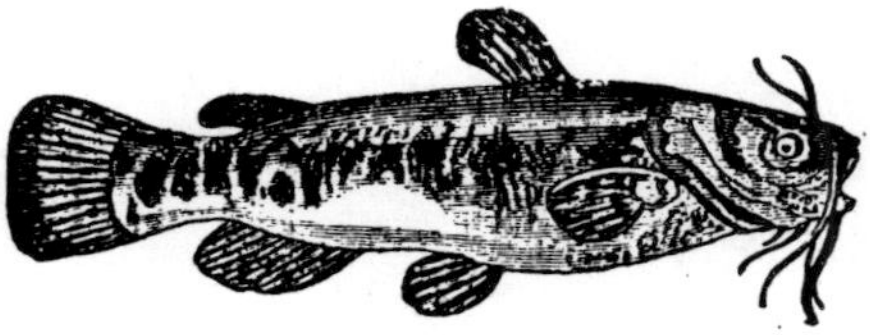

The way they do this is with two of the most sinister looking things imaginable hanging from the sides of its mouth. Actually, there are several of these appendages around the mouth of the bull head, but it's the two main ones that fascinate people. Most folks think that anything that sinister has got to somehow hurt like all get out.

So, while a person is thus carefully avoiding those long scary looking things the bullhead has around his mouth, he'll give a quick thrust of its head and drive a fin spike right into your hand.

Now, there is something that can hurt like all get out!

CHAPTER VII

RATCHETY COONS

These lovable little piles of fur and face masks are not a separate and complete species in and of themselves. They are simply biologically remote appendages to the sweet corn plant. Their function in the grand scheme of things is to sneak in and pick the sweet corn in the night just hours before you plan to.

The ability of the ratchety coon to time this operation goes far beyond awe-inspiring, straight to breathless.

The ratchety coon's role in things seems to be limited to the sweet corn thing, and it littering our highways with its ~~carcass~~ ~~carracas~~ ~~carrcas~~ body.

CHAPTER VIII

DOUBLE-BREASTED MATTRESS FLAPPER

One of Illinois' most attractive creatures is unique in that all members of the species are female. This lovely little critter is the Double-Breasted Mattress Flapper. Its scientific name is The Bare-Bottomed Sleep Stealer.

There has been remarkably little research done on the feeling, shopping, or migratory habits of this creature since your basic research scientist tends to get kind of giddy and silly in the presence of this creature. All of this proves there is hope for them yet.

That is, I mean, hope for the research scientists . . . . . that is.

CHAPTER IX

THREE-BEAN SALAD

There are schools of opinion about the qualifications of Three-Bean Salad for inclusion in the defination of animal life.

But, distracting voices aside, Three-Bean Salad is undoubtedly a critter. In fact, a careful study of world famous Dr. Julian's defination of animal life shows that Three-Bean Salad qualifies.

While the scientific background can be rather voluminous, Dr. Julian's defination can ultimately be expressed in three rather simple and straightforward sentences. These are:

(1) A living thing has cellular morophological form.

(2) A living thing undergoes self-actuating regeneration.

(3) A living thing contains beans.

And those who would deny that Three-Bean Salad meets those qualifications just simply doesn't know their Three-Bean Salad.

And, Three-Bean Salad is certainly an Illinois critter since its range encompasses the entire state. It is as much at home here in the Land of Lincoln in Shawneetown down in Gallatin County as it is way up in Oregon in Ogle County.

Like most of Illinois critters, Three-Bean Salad in not migratory. That is, it stays here year 'round. You are as likely to spot a Three-Bean Salad at a Christmas potluck as you are a Fourth of July picnic.

Much remains to be learned about the reproductive habits of Three-Bean Salad. There is even some doubt if it does reproduce or not. Perhaps, like the mule, it simply does not. It has been noted that Three-Bean Salad kept in the darkness seems to have no more propensity for reproduction than that kept in the light. Maybe it just doesn't care about reproduction.

We don't know what the mule's attitude is about reproduction either.

CHAPTER X

THE FUZZY CHINNED LOOP-SCOOPER

The Fuzzy Chinned Loop-Scooper is unique in that individuals within this species are almost all of the same sex; male.

While considerable research needs to be done to fully understand the bizarre behavior patterns of this creature, it appears to be simply a case of their indulging in some sort of pre-adult

ritual. This ritual involving patterns of movement and sound consists of traveling around and around in loops of urban streets. In fact, in rural areas those members of this species will migrate nightly to urban ones that have such loops.

While scooping the loop is almost always done in automobiles, other vehicles have been used on ocassion. A favorite among the alternatives is the motorcycle.

It has been suggested that since this behavior pattern appears to be so inherently mindless, that it apparently has something to do with a mating ritual.

What few females there are, are often found perched on automobile fenders. Apparently that is some sort of pre-mating posturing.

CHAPTER XI

THE ILLINOIS HIGH SCHOOL BIOLOGY CLASS DISSECTION FROG

Now, I don't know how many of you folks have ever given a whole lot of thought to what a career of being an Illinois High School Biology Class Dissection Frog must be like, but you can rest assured that the dying and the embalming in formalhyde has gotta be

"I know this isn't a good likeness of a frog, but it's the best I could do this morning."

as good as it gets. It surely must be all downhill from there.

First thing, as soon as Anatomy 101, 5th Hr., starts, you are faced with Trial #1. Some hairy hand that just got done fondling a peanut butter smeared piece of Pizza at lunch grabs you up to scare some junior girl whose falsetto screams will curdle dry air. And, that's only the beginning.

After that, it's the losing of parts. These losses are best not gone into in any detail, but they involve losses that vary from the embarrasing to the terrible.

Then, just as you are sure you are going to OD on all those "Oh, Grosses!", the bell rings. Then you're left out to dry until some solemn-faced young biology teacher appears and stuffs your innards back in so you'll be all set for trial #2 when 7th hour anatomy class starts.

How's that for a fun day?

CHAPTER XII

SEVENTEEN-YEAR LOCUST

While we're talking about Illinois animals with problems, we might as well talk about the Seventeen-Year Locust.

While this critter doesn't have to listen to all those "Oh, Grosses" and doesn't have to put up with losing all those parts he'd kind of grown fond of, he still does have a problem. He does have his own cross to bear.

It isn't enough that all he gets to do is to to hang around underground for seventeen years and then gets to cling to a tree trunk for a couple of days. The worst part is when some joker confuses him with that Seven-Year-Itch mentioned earlier.

I mean, there you are, enjoying hanging out on that tree trunk after seventeen years underground. It isn't what you'd call heart-stopping exciting, but it beats the first tour of duty. Then some guy comes up to you and kind of pokes you in the thorax with his elbow.

Then, comes the worst part when he asks you, with a wink, that same old question you'd heard before.

"How's it goin', Buddy? How's the old Seven-Year-Itch?"

I mean, right then is when you suddenly realize that hanging out on a tree truck is really pretty tame, after all.

CHAPTER XIII

FAT-OF-THE-LAND

Shown below is Illinois' offical Fat-Of-The-Land. This is what fellows convince their wives is free food, just for the shooting.

Technically, of course, it's really a ring-necked pheasant and is usually called that.

It is, on the other hand, called Fat-Of-The-Land when a man convinces his bride that he can shoot them for some Sunday dinners.

This little ruse works only if the blushing bride doesn't know anything about hunting, of course. That's why most hunters, or would-be hunters marry girls who don't have a father or brothers who hunt.

"Why, all I have to do, Honey, is to buy a box of shells, and borrow a gun off one of the fellows at work. With a dozen shells, I ought to be able to bag ten of those Fat-Of-The-Lands."

At this point the little honey mentally makes note of the price tags on those pretty fryers in the supermarket. She can understand the wisdom of her husband buying that box of shells. It would be a real good investment and would be their ticket to some free food, living off the fat of the land, don't ya know.

There are, of course, not enough pages in this book to list all the costs of hunting that Mr. Livin'-Off-The-Fat-Of-The-Land will eventually encounter as he takes up the practice of hunting on the prairies of Illinois.

But what with those shells plus a license, then his own gun, the dogs, and their care and feeding, it all adds up. Then toss in a few thousand gallons of gas, several broken suspension systems on the pickup truck he has to have, a gun rack, lots of beer, equipment and supplies.

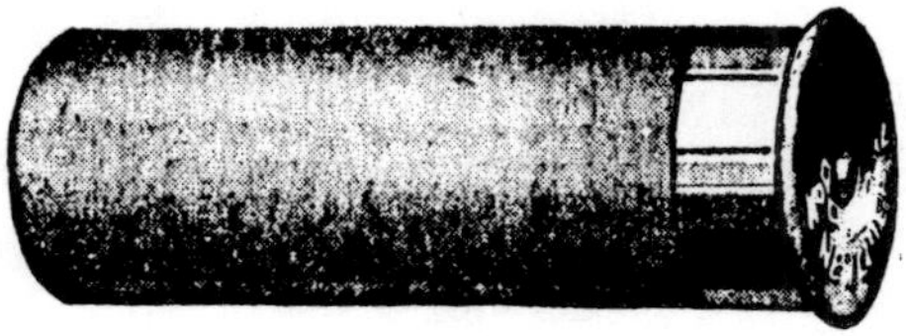

Be sure to include the magazines, reloading equipment, the hundred and fifty dollars boots and the special sun glasses.

When you do all that, the cost per pound of the Fat-Of-The-Lands that he ends up getting is a modest $867.43.

That's a bit more that the fryer there in the supermarket, but the fun you can have in the dentist's chair as a result of biting down on the lead shot more than makes up for the price difference between Fat-Of-The-Lands and mere chickens.

Eventually, of course, wifey gets wise to the economics of hunting Fats-Of-The-Lands, but by then the investment is made and it's too late.

CHAPTER XIV

MONARCH BUTTERFLIES

One of the most insoluable mysteries about the animals in Illinois is about the Monarch butterfly.

The mystery, of course, is just how, they manage that migration they go through every year.

Every fall those big beautiful butterflies migrate south to somewhere in South America, or maybe it's Africa. It doesn't make a whole lot of difference if you're not going with them. Whereever it is, it's a long, long way from Peoria.

Now, flying thousands and thousands of miles is no great big deal. A lot of critters do that. There are lots of birds that go that far when they migrate.

The mystery is just how those Monarch butterflies manage to do it. If you ever watch a bird migrating, they get right in there and do it. They'll heave hove hive come into sight over past your neighbor's barn to the north; and as you watch, they will pass over your place, and in a minute or two, they'll be out of sight to the south. That's the way a bird'll do it. It's kind of the aeronatical equivalent of pickin' 'em up and settin' 'em down.

You do that for twelve or fifteen hours a day, and you've got a few miles under your belt. Then you do that day after day and you'd better start learning to tweet in Spanish, 'cause you're gonna be down there.

But, compare that to a Monarch butterfly as they migrate south in the fall. Just pick out one of those little suckers and keep an eye on him to see how he manages for a half hour or so.

You'll see pretty quickly that they just kind of bat themselves around in the air, going all six directions, and everywhere in between. They'll flutter straight up for a while for no apparent purpose at all.

They'll head south for a couple of feet, then veer off to the northeast. Then, like as not, they'll light on a clothesline wire to watch it rust for a while.

After about half an hour, they'll be a few feet farther south than when you started watching them. The whole thing is like watching a government employee at work.

Yet, after about a week or so ofo this, they've gone a jillion miles to roost in some exotic tree a scant few hundred miles from Anarticia.

I'm just not real sure I understand all that.

CHAPTER XV

THE CHEERLEADER

One of Illinois' most unappreciated critters is the cheerleader. The cheerleader is an exception to the general rule that any given animal can have only one special thing going for it, you know.

The Cheetah, for example, is one fast critter but looks like an aneroxic alley cat.

The Koahla Bear is cute as all get out but is dumb as a corner post. Endless examples can be ~~sited~~ ~~cited~~ ~~sitted~~ told about to prove that rule.

The cheerleader, however, has two major things going for her. First of all, she is pleasant appearing, being clean of limb, and of cheerful contenance. And those are just the characteristics you'd want to tell your mother about.

Second of all, the cheerleader is the only known critter in Illinois that is approximately 8.4 times as smart as the other critters immediately around it.

Think about it. Here's this situation wherein there's all those critters out on a playing field or floor. The larger ones are all decked out in some sort of

special plummage, throwing, batting, or hitting some sort of spherical oblong ritual device. They propel that device back and forth, simultaneously poking, hitting, falling on, and generally playing havoc on each other's bodies. Those folks are out there sowing the seeds of brain concussions, fractured limbs, and arthritis.

While all that's going on, those cheerleaders aren't having any part of such foolishness other than to urge the lesser species on to ever greater acts of violence and mayhem. Those clever little cheerleaders are simply leaping and jumping around, showing off their pretty little . . . . outfits. Now, you tell me which of those two species has their act together the more. In fact, let's change that 8.4 to 11.3.

CHAPTER XVI

THE NUMBERED COW

While most of us have little need to be introduced to the Illinois cow, there is one breed of cow that isn't seen much anymore. Unlike the Hereford or the Shorthorn, this one just isn't around any more. It fell victim to the fickleness of technology.

This rather specialized breed was one that was developed to serve the needs of our growing nation and the need to have the capability of showing the populace various things in publications such as encyclopedia, dictionaries, textbooks, etc. Thus was developed The Numbered Cow.

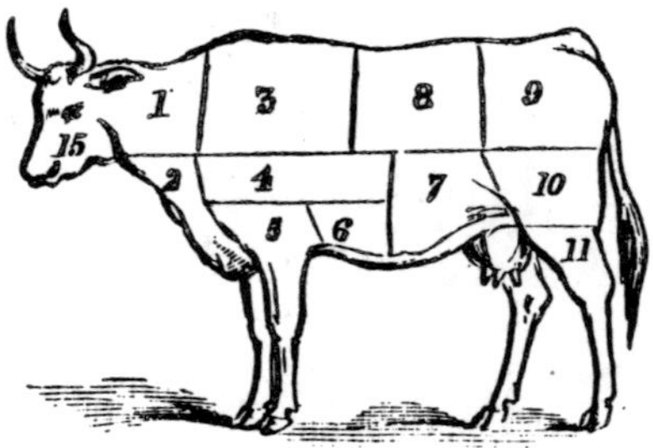

The Numbered Cow was especially bred to provide models for illustrations in encyclopedia, dictionaries, textbooks, etc. Through selection breeding The Numbered Cow Breeding Association was able to come up with a cow with numbers clearly and accurately portrayed on its side for this rather specialized use.

The fate of The Numbered Cow Breeding Association was pretty well sealed when modern technology enabled illustrators to take an outline of an ordinary cow, draw in some lines and numbers and make their own numbered illustrations of a cow. When that happened, the prices of The Numbered Cow first plunged, then the market just evaporated altogether.

The Numbered Cow is now, in fact, so rare that many people have never even seen one of the critters, much less an entire herd such that frolicked on the prairies of Illinois in days gone by.

CHAPTER XVII

BUNNIES

If you think, Dear Reader, that you are safe here in Illinois from the ravages of earthquakes, think again. Just over the border to the west in Missouri is the epicenter of the most dangerous geologic fault in the entire North American continent.

And just because it's been a century and a half since the New Madrid fault cut loose with any seriousness, don't think it can't happen again. In fact, this rather long period of inactivity is cause for anything but complacency. When it eventually lets go again, it ought to let go with a real vengence.

I also would ask you to think about the proposal by some serious scientists that we fritter away our tax dollars by funding research on earthquake prediction by the observation of the behavior of wild animals. This is, of course, the sort of thing that has been done in China for several years now.

Apparently, the Chinese experience has been that rabbits are particularly good at picking up those tiny movements from deep within the earth and that rabbits could well be used to warn us of an

impending earthquake. The theory is that they would respond to those tiny movements that preceded those building-shaking-down ones.

Before you get too excited about that idea, I would call your attention to a couple of relevant facts about those cuddly little bunnies you should remember.

One of these is that the expression "hare-brained" didn't just happen.

The other is that these are the same little critters that will lurk along the side of the highway at night, not feeling moved to jump out onto the concrete until your headlights are right on 'em.

What I want to know is just how are we to depend on these little critters to pick up on those tiny underground vibrations when they can't see or hear an eighteen-ton truck coming at 'em on the highway?

Scientists are alright to have around, but you do have to keep an eye on them.

CHAPTER XVIII

THE OPTIC

Speaking of selective breeding, many of our critters here in Illinois have been selectively bred for one purpose or another. In some critters, this has been done by man for his own purposes. In other cases it has ocurred naturally according to the dictates of nature.

The classic example of natural selective breeding here in Illinois is one that is often used in biology classes to illustrate the fact that a critter can become so sophisticated that it cannot survive.

It can concentrate so much on developing only one or two capabilities that other functions get neglected. This can ultimately be to the dire disadvantage of the critter.

So it was right here in Illinois with the Illinois River Valley Optic.

The Illinois River Valley Optic, being a nocturnal animal, found it necessary to do its hunting and foraging at night. This, in turn, required it having night vision considerably superior to that of either its quarry or its predatory enemies.

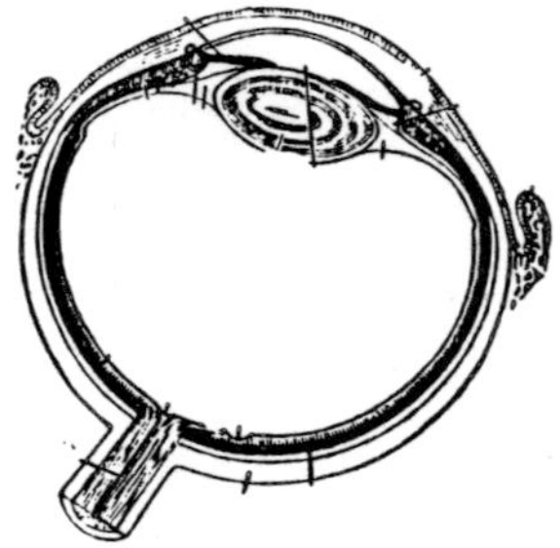

What happened was that the normal and reasonable development of night vision just simply got out of hand. The Optic also developed excellent perpherical vision and excellent vision for things very close and small. It also developed equally good vision for things far in the distance.

Before Mother Nature could get things sort of balanced out, the Optic developed its optical capabilities so well that it ended up being nothing but one big old eyeball.

Even the common mole with its poor vision but with legs and all proved to be a formidable predator on the Optic because the Optic went too far and found himself without either protection from its enemies or means of locomotion.

The Optic found out pretty quickly that if a critter's comprised entirely of an eyeball, even if its a good one, it is at a decided disadvantage when it comes to foraging or escaping.

Even before the last of the Optics disappeared from the prairies of Illinois, its fate was almost predetermined. It simply wasn't destined to make it in a world of animals that had legs, fins, feathers, or whatever.

CHAPTER XIX

THE STRONG-TOED BAT

Always a little mysterious and always just a little bit scary are the various bats that make their home here in Illinois.

And the one that is probably the most interesting of all is the one whose scientific name is Strong-Toed Bat. Those words come from the Greek "Strong-Toed" meaning; strong-toed.

These particular bats are unique in that they occur in three totally different colors. They come in black, brown, and a sort of milky grey. It has been suggested that the reason for these various colors is to provide protective coloration with the various colors of rock in the caves these bats live in here in Illinois.

Less easy to understand is why these bats rest the way they do. Instead of hanging from things they set on top of twigs and other roosts. They perch on top of the roost like a bird will.

BAT

Strangely enough, they wrap their toes around the under part of the perch instead of the top. One can certainly understand why they would be named Strong-Toed Bats, of course.

CHAPTER XX

GNATS

Time will prove that this volume is much more than simply a ready-reference guide to the critters here in Illinois. Time will prove that it was first on these pages that a solution to the gnat problem was first proposed that would eventually result in the end of those pesky little critters.

The gnat problem is, of course, the very existence of those useless creatures. It can be easily seen that gnats have no socially redeeming characteristics and should be dispatched to whereever sirens, sea monsters, and the passenger pigeon went.

I mean you can't really blame a mosquito for taking a bite out of you or a fox for doing in a chicken. It is necessary for those animals to do that in order to stay alive.

But what kind of survial value is there in buzzing around in people's faces all day? It doesn't serve anyone's interests and just shows that gnats are gnot fit to have a gniche in this world.

A solution to the problem is to adapt to the nature of things and to try to raise gnats. We could put together large gnat ranches. Pamphlets could be printed up by some government office that would go into great detail on the care and feeding of gnats. Gnat care seminars could be held on all of our campus's here in Illinois.

In other words, we could make an all-out effort to culivate the well being of gnats just like we do cattle and hogs.

And, like in the case of cattle and hogs, all sorts of things would get in the way.

Just about the time we got going good, some rare

gnat ailment would sneak into the United States in a crate of imported avocados or something like that. Immediately the disease would spread like wildfire throughout the United States. We would find the gnat population wiped out. Ipso facto, E Pluribus Unium, we'd be shed of all those gnats.

Just remember, you heard it here first.

CHAPTER XXI

SIRENS

A lot of folks who don't know any better think of Illinois as being just a tad wetter than Kansas or Oklahoma.

The truth is, of course, that Illinois is a seafaring state what with Lake Michigan, and the Mississippi, Rock, Ohio, and the Illinois Rivers.

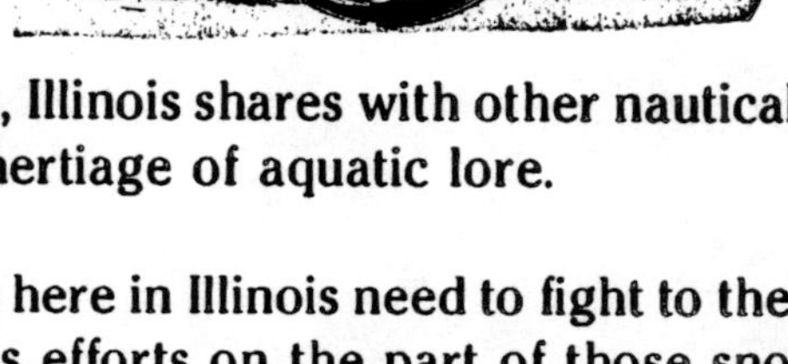

As such, Illinois shares with other nautical places, a rich hertiage of aquatic lore.

And we here in Illinois need to fight to the last the insidious efforts on the part of those snotty professors in their prestigous eastern colleges and universities to rob us of our heritage.

One of the more unconscionable acts of cultural pillage that those people have tried to do is to claim that the siren was an ancient Greek critter.

These false legends have it that Odysseus was warned by Circe not to pay heed to the sirens with their enticing songs of love and passion there from the rocks off the coast of Greece.

All this was alleged to have been recited in the works of Homer. Every serious student of ancient history knows, however, that it wasn't that way.

It was actually Howard Minden from Pekin telling his neighbor Jeff Kroner about the sirens in the Illinois River.

It seems that Jeff tended to have a fascination for a well-turned ankle and was somewhat of a push-over when it came to siren songs of love and

passion. He was also a commerical fisherman on the Illinois and did a little farming outside of Pekin between times.

It also seems that Howard's warnings about those sirens that set up on sandbars in the Illinois fell on deaf ears. The story goes that Jeff lost four good john boats to the snags these in the Illinois before he figured out the best way to handle the situation was to tie up on upriver a ways and walk down to the sandbar.

Sirens are, of course, pretty much regulation type constructed honeys except for their feet that are the claws of huge birds. Such a foot tends to kind of mess up what would have otherwise been a well-turned andle. But Jeff soon got over that. He discovered that honeys are a whole lot more than well-turned ankles anyway.

CHAPTER XXII

SHELF BRACKETS

Now, I'd be the first one to admit that to include a shelf bracket that's in the shape of a voluptuous honey in the animal kingdom is pushing a point a bit. I'm not saying that such a device technically qualifies as an animal. But, after all, I'm the one writin' this book, and if I can't take a liberty or two with it, who can?

I do think, though, that while the cutural and sociological impact of shelf bracket design doesn't always strike a person as being really super important, we can't let a thing like that sneak up on us. We need to be alert to the possibility that things can change slowly and insidiously. You know, today the voluptuous honey shelf bracket, and tomorrow some other species will be lost.

Next thing you know, our children will inherit a world without the rich diversity of life forms that makes it all worthwhile.

Do you know what is happening in shelf bracket design in your community tonight?

CHAPTER XXIII

THE AMOEBA

The most ubiquitous of all the animals in Illinois is undoubtedly the amoeba. That little sucker is found in every city, hamlet, farm, and body of water in the state.

It is found under, on and inside of our fair state's citizens. The amoeba was here long before either the white man or the Indian. It was here before The Great Flood and after the water subsided. The amoeba survived the earthquake of 1831 and the indiscretion of Mrs. O'Leary's cow.

It is this author's opinion that we need to change our state animal to the amoeba (that is, unless the cheerleader would accept that office). We would thus recognize the importance of that little critter to our fair state.

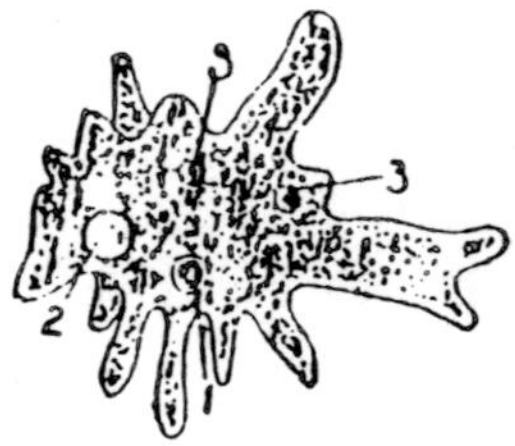

In terms of qualifications for designation as the state animal, the amoeba simply goes unchalleng-ed. Its widespread habitat is certainly a considera-tion as well as its asexual (think of all the problems that would solve) reproduction. The amoeba doesn't mess in peoples' yards and doesn't bite, kick, and sting.

CHAPTER XXIV

GEORGE CANE'S SEA MONSTER

The account of George Cane's sea monster turned out to be false but it caused such a stir at the time down near Keithsburg where George had his experience that I'm going to report it on these pages as it was alleged to have happened. I am also including a reproduction of the photograph taken by young George who was a deckhand on the T.S. Thomas.

The T.S. Thomas was a side wheeler that plied the Mississippi, carrying both cargo and passengers.

On the night of this incident, George was on watch as the T.S. Thomas laid at anchor just south of Keithsburg, waiting for the coming of dawn before continuing on downriver.

George swore later that he was totally sober and alert when he first heard some splashing around out in the water near a snag off the port side.

Even before he could walk over to the railing to investigate, however, he was shocked to see a scaly arm or leg of some kind coming up over the side, and a creature was pulling itself up as huge globs of slime and debris fell from its scale-covered body.

George was, of course, terrified almost to the point of being struck motionless. But he was able to dart into his cabin and come out with his camera.

George explained later that the light wasn't the best, but there was a full moon and the light from the forward flare helped a lot. He told of how he hurriedly pointed the camera in the general direction of the creature and flipped the shutter.

His account as recorded later in a local newspaper told of how the creature stood there on the deck surveying George, and how the young man had the distinct impression he was being sized up as to how he'd do as a meal.

It didn't take George long to dash down along the starboard railing to rouse the Captain. Apparently George's frantic blabbering about this huge monster standing on the deck there midship at the port side was enough to galvanize Captain Slade to grab up his gun and follow the lad.

When the pair got back to the spot but a minute or two later, the creature was gone. There proved to be no evidence of it ever having been there or anywhere else on the T.S. Thomas that night. Captain Slade even dispatched a reluctant crew member over the side to inspect the hull for scratch marks or other evidence of the animal having crawled up onto the deck.

George's insistence on the truth of what he saw that night was met with only the jeers and taunts of his shipmates. He was teased for his overactive immagination that night.

Captain Slade gave strict orders to all the crew that they were not to talk to any of the passengers about the incident. Even if it was a false report, he didn't want it to be known that the whole incident had been said to happen. He knew that even such a rumor could be bad for business.

I the face of that continual teasing by the rest of the crew, even George begin to wonder if he might just have dozed off that night and dreamt of seeing that critter and taking its picture.

But when he got those pictures in his camera developed, it was a whole 'nother story.

For there it was! It was a scaly looking critter with what appeared to be some sort of seaweed attached to its belly. Its large body was supported by four almost spindly legs, each terminating with what appeared to be a hoof-like foot.

Folks were surprised to see that the animal appeared to have no head, but it did have what looked like a place like a turtle had into which the head could be retracted.

George said that things had happened so quickly and he was so scared that he hadn't really a good recollection of what the creature's head looked like.

With or without a head out where folks could see it, that animal was definitely one weird-looking beast and George was considered to be something of a local hero for having had the experience of seeing it and having had the presense of mind to fetch his camera to get that picture.

George's picture of that critter showed up in the local newspaper for several days, giving all the folks in the area a chance to see it and to speculate as to what kind of unspeakable beast had boarded the T.S. Thomas that night.

And, best of all, George didn't have to take any more guff from the other crew-members about his overactive immagination.

Rather than the news of the sea monster discouraging passengers from taking the T.S. Thomas, the fame of George's sea monster attracted the curious and Captain Slade had all the business he could handle.

Meanwhile, George found life pretty easy since all he was asked to do was to lounge around with the passengers, eating and drinking well as he recited over and over again the details of that night he was on watch and saw the great sea monster.

George was even asked to bring out the camera he took that famous photograph with. All in all, it was the kind of life that George had always secretly longed for. So a person couldn't blame George for how he felt the day it all fell apart. After all, George could have built a nice easy life and lucrative career out of his experience, but a discovery by his cabin mate, Will Clarence, one day dashed all that.

Not only was it very unfortunate for George that Will made that discovery, but then Will had to use the poor judgment he did in bursting right out there with all those passengers around to tell about it.

And, to make it even worse, there was a reporter from a newspaper right there in that group that overheard Will's conversation.

"Why, lookie here, George! Lookie at what I discovered whilst I was studyin' your sea monster picture."

"What's that, Will?"

"We bin lookin' at this picture upside down all the time. That ain't no sea monster. That's jus' me layin' in my hammock down in our cabin. 'Member the day you tuck my picture when I was layin' in my hammock?"

With trembling hand, George took that photograph from his cabinmate's hand to look at it. It took but a second or two for George to realize that what Will was saying was true.

How could it be? How could he have gotten that

picture and looked at it upside down? Where was the photograph of the sea monster that had offered him a ticket to a whole new way of life? Could it have been one of those that didn't turn out?

The reporter was on George in a minute, snatching the photograph out of the poor lad's hand, sensing a big story was in the making.

And he was right. The big story was, of course, that the first big story was only a story, and there hadn't been any sea monster aboard the T.S. Thomas. Even before George could regain his composure, the man from the newspaper was hurrying down the deck to find his cabin and his typewriter.

It wasn't long until people all over the T.S. Thomas were all turning their newspapers upside down to examine anew the photograph of George's sea monster.

Sure enough, the monster was nothing but a common seaman's hammock. The "seaweed" was merely blankets hanging over the side.

Those "feet" on the deck proved to be nothing but hangers from the ceiling. Even George's cabinmate's face could plainly be seen, a mere detail that had been overlooked when the photo was viewed upside down.

George's dream of a life of fame and luxury dissolved overnight and he had to go back to work.

CHAPTER XXV

THE RED-BEAKED BESPANGED PUSHOVER

At first blush one might tend to be confused between this creature of attractive, if limited, plummage with the more prevalent Double-Breasted Mattress Flapper, if one were to consider only their modes of behavior.

The similarity of behavior is quite superficial, however. What the Red-Beaked Bespangled Pushover does for economic reasons, the Double-Breasted Mattress Flapper does for entertainment.

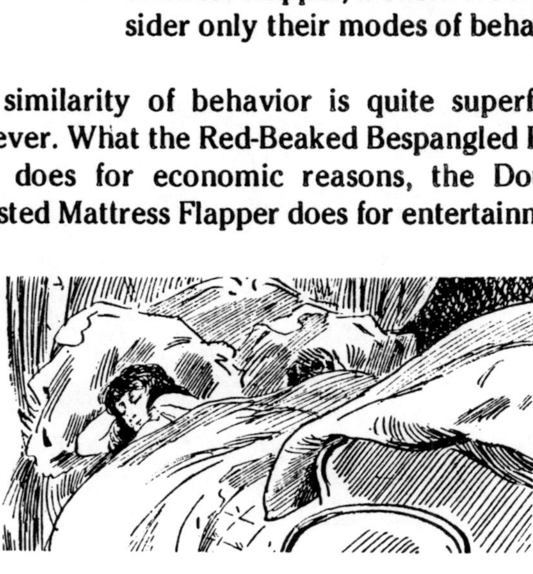

Unlike the Flapper, the Pushover is almost exclusively a nocturnal critter, emerging from its nest in the early evening and likely to exhibit considerable activity throughout most of the hours of darkness.

Like the puff adder or the Spiney Blow Fish, the Red-Beaked Bespangled Pushover has the capability of exaggerating certain of her anatomical features to effect a cosmetic change in appearance.

Current theory is that both the puff adder and the blowfish do this when frightened or threatened, giving them a vicious appearance to repel others of their species or some predatory species.

The effect of the Pushover's anatomical changes,

however, seems to have the opposite effect, and careful examination of this spiney blow fish on this page and the pushover on page 96 tends to confirm this theory. Most people will agree that the Red-Beaked Bespangled Pushover looks considerably better than the Spiney Blowfish.

EPILOGUE

So we find Illinois' fauna to be a varied lot. Some are old friends like the *Bullhead* and the *Monarch Butterfly.*

Others like the *Mayfly* and *Gnats* are just plain old.

Then there are those subtle species like the *Fat-of-The-Land* and the *Double-Breasted Mattress Flapper.* These can get pretty exciting.

And, since the last definitive work on this subject, new discoveries have been made. The *Optic,* and *Fuzzy-Chinned Loop Scooper* come to mind as does the *Sailer Cat* and *Three-Bean Salad.*

We hope you enjoy this book so much that you'll be moved to run out and buy half a dozen copies for gifts.

If you have enjoyed this book, perhaps you would enjoy others from Quixote Press.

GHOSTS OF THE MISSISSIPPI RIVER
Mpls. to Dubuque by Bruce Carlson ....paperback $9.95

GHOSTS OF THE MISSISSIPPI RIVER
Dubuque to Keokuk by Bruce Carlson .....paperback $9.95

GHOSTS OF THE MISSISSIPPI RIVER
Keokuk to St. Louis by Bruce Carlson .....paperback $9.95

JACK KING VS. DETECTIVE MAC KENZIE
by Netha Bell ....paperback $9.95

HOW TO TALK MIDWESTERN
by Robert Thomas ....paperback $7.95

GHOSTS OF DES MOINES COUNTY, IOWA
by Bruce Carlson ....hardback $12.00

GHOSTS OF SCOTT COUNTY, IOWA
by Bruce Carlson ....hardback $12.95

GHOSTS OF ROCK ISLAND COUNTY, ILLINOIS
by Bruce Carlson ....hardback $12.95

GHOSTS OF THE AMANA COLONIES
by Lori Erickson ....paperback $9.95

GHOSTS OF NORTHEAST IOWA
by Ruth Hein and Vicky Hinsenbrock .....paperback $9.95

GHOSTS OF POLK COUNTY, IOWA
by Tom Welch...........................paperback $9.95

GHOSTS OF THE IOWA GREAT LAKES
by Bruce Carlson.......................paperback $9.95

MEMOIRS OF A DAKOTA HUNTER
by Gary Scholl.........................paperback $9.95

LOST AND BURIED TREASURE ALONG THE MISSISSIPPI
by Gary Scholl and Netha Bell ....paperback $9.95

(Continued on Next Page)

MISSISSIPPI RIVER PO' FOLK
by Pat Wallace ....paperback $9.95

STRANGE FOLKS ALONG THE MISSISSIPPI
by Pat Wallace.........................paperback $9.95

THE VANISHING OUTHOUSE OF IOWA
by Bruce Carlson......................paperback $9.95

THE VANISHING OUTHOUSE OF ILLINOIS
by Bruce Carlson......................paperback $9.95

THE VANISHING OUTHOUSE OF MINNESOTA
by Bruce Carlson......................paperback $9.95

THE VANISHING OUTHOUSE OF WISCONSIN
by Bruce Carlson......................paperback $9.95

MISSISSIPPI RIVER COOKIN' BOOK
by Bruce Carlson ....paperback $11.95

IOWA'S ROAD KILL COOKBOOK
by Bruce Carlson......................paperback $7.95

HITCH HIKING THE UPPER MIDWEST
by Bruce Carlson ....paperback $7.95

IOWA, THE LAND BETWEEN THE VOWELS
by Bruce Carlson......................paperback $9.95
(Farm Boy Stories From the Early 1900's)

GHOSTS OF SOUTHWEST MINNESOTA
by Ruth Hein ....paperback $9.95

GHOSTS OF THE COAST OF MAINE
by Carole Olivieri Schulte ....paperback $9.95

ME 'N WESLEY
by Bruce Carlson......................paperback $9.95
(Stories about the homemade toys that farm children made and played with around the turn of the century.)

SOUTH DAKOTA ROAD KILL COOKBOOK
by Bruce Carlson .... paperback $7.95

GHOSTS OF THE BLACK HILLS
by Tom Welch .... paperback $9.95

Some Pretty Tame, But Kinda Funny Stories About Early DAKOTA LADIES-OF-THE-EVENING
by Tom Welch .... paperback $9.95

Some Pretty Tame, But Kinda Funny Stories About Early IOWA LADIES-OF-THE EVENING
by Bruce Carlson .... paperback $9.95

Some Pretty Tame, But Kinda Funny Stories About Early ILLINOIS LADIES-OF-THE-EVENING
by Bruce Carlson .... paperback $9.95

Some Pretty Tame, But Kinda Funny Stories About Early MINNESOTA LADIES-OF-THE-EVENING
by Bruce Carlson .... paperback $9.95

Some Pretty Tame, But Kinda Funny Stories About Early WISCONSIN LADIES-OF-THE-EVENING
by Bruce Carlson .... paperback $9.95

Some Pretty Tame, But Kinda Funny Stories About Early MISSOURI LADIES-OF-THE-EVENING
by Bruce Carlson .... paperback $9.95

THE DAKOTA'S VANISHING OUTHOUSE
Bruce Carlson .... paperback $9.95

ILLINOIS' ROAD KILL COOKBOOK
by Bruce Carlson .... paperback $7.95

OLD IOWA HOUSES, YOUNG LOVES
by Bruce Carlson .... paperback $9.95
(Stores about old houses in Iowa and young loves they have known.)

TERROR IN THE BLACK HILLS

by Dick Kennedy .... paperback $9.95

IOWA'S EARLY HOME REMEDIES

by 26 students at Wapello Elem. School ... paperback $9.95

GHOSTS OF DOOR COUNTY, WISCONSIN

by Geri Rider .... paperback $9.95

TALES OF HACKETT'S CREEK

(1940's Mississippi River Kids)

by Dan Titus .... paperback $9.95

A FIELD GUIDE TO IOWA'S CRITTERS

by Bruce Carlson .... paperback $7.95

A FIELD GUIDE TO ILLINOIS' CRITTERS

by Bruce Carlson .... paperback $7.95

A FIELD GUIDE TO MISSOURI'S CRITTERS

by Bruce Carlson .... paperback $7.95

MISSOURI'S ROADKILL COOKBOOK

by Bruce Carlson .... paperback $7.95

MINNESOTA'S ROADKILL COOKBOOK

by Bruce Carlson .... paperback $7.95

EARLY IOWA SCHOOLS

by C. Johnston .... paperback $9.95

Need a Gift?

For

• Shower • Birthday • Mother's Day •
• Anniversary • Christmas •

Turn Page For Order Form
(Order Now While Supply Lasts!)

TO ORDER COPIES OF

A FIELD GUIDE TO ILLINOIS CRITTERS

Please send me________copies of **A Field Guide to Illinois Critters** at $7.95 each. (Make checks payable to **QUIXOTE PRESS.**)

Name ____________________________________

Street ____________________________________

City ____________ State ________ Zip Code ____

SEND ORDERS TO:

QUIXOTE PRESS
R.R. #4, Box 33B
Blvd. Station
Sioux City, Iowa 51109

TO ORDER COPIES OF

A FIELD GUIDE TO ILLINOIS CRITTERS

Please send me________copies of **A Field Guide to Illinois Critters** at $7.95 each. (Make checks payable to **QUIXOTE PRESS.**)

Name ____________________________________

Street ____________________________________

City ____________ State ________ Zip Code ____

SEND ORDERS TO:

QUIXOTE PRESS
R.R. #4, Box 33B
Blvd. Station
Sioux City, Iowa 51109

Index

Anchovies 64
Axle Grease 15,42

.... Oh, what the heck. You don't really need an index for this book. Compiling an index can really be a pain in the neck and I'm not sure I'm up to it this morning.

Thanks for buying this book!
I needed the money.

.... *Bruce Carlson*